BR
ELECTRIC MULTIPLE UNITS

Brian Hardy

D. BRADFORD BARTON LTD

© copyright D. Bradford Barton 1980   9085/3BL   ISBN 0 85153 368 X

printed in Great Britain by Lovell Baines Print Ltd, Newbury, Berkshire,

for the publishers

D. BRADFORD BARTON LTD · **Trethellan House** · **Truro** · **Cornwall** · **England**

Class 304/1 a.c. Unit 004 leading at Stretford, on an Altrincham to Alderley Edge working.

introduction

It is rather appropriate that this book should be compiled in 1979, the year which celebrates one hundred years of electric traction on our railways. Over the years, two separate systems of electrification have been developed; one being current collection by conductor rails at about running rail level, and later the other, developed current collection by overhead wires. This volume is designed to illustrate electric multiple unit rolling stock in use in the post-Nationalisation period, showing the different design variations of the rolling stock on the various systems.

It should be pointed out, however, that the Southern electric multiple unit types are not catered for herein, as these are covered fully in a companion Bradford Barton publication ('Southern Electrics', published 1975) solely devoted to the subject.

At the present time, electric railway systems in this country are widely expanding. The first of the new rolling stock for the Argyle line in Glasgow has been delivered and has started commissioning trials, and delivery is imminent of new Class 508 stock for Southern Region. The latest electrification scheme currently being carried out is that between Bedford and London, due for completion in 1982, and the Tyne & Wear Metro in Newcastle, which is taking over parts of B.R. line that were originally electrified into the 1960s, but were subsequently dieselised.

The current trend towards electrification can surely be expected to expand in future years. With the continuing oil crisis, some of the 'talked-about' schemes may be pushed for implementation sooner than envisaged.

At this point, I would like to thank all the people who responded to my request for photographs, for without their help the range and variety of subjects covered would not have been possible.

LIVERPOOL OVERHEAD RAILWAY
525/550V d.c. 3rd rail

The L.O.R. was opened in March 1893 and was the first electric railway in Britain to use e.m.u. rolling stock, picking up current from a centre third rail. This was altered to outside one of the running rails in 1905 when inter-running started with the L. & Y. electric lines. The L.O.R. did not become part of B.R. in 1948 when the railways were nationalised but continued to be independently owned. Because the Company was unable to meet the cost of renewing the overhead structure, the line closed permanently on December 1956. The original rolling stock dated back to 1893-7 and was built by Brown Marshalls, who became part of what is now Metro-Cammell. Seen at Herculaneum Dock is 1896-built car No.41 leading on 31 December 1955.

L.Collings

Seen arriving at Herculaneum Dock approaching from Dingle is a three-car train with 1893 motor car No.25 leading. This was one of ten motor cars rebuilt in 1902-3 having the body width increased from 8' 6" to 9' 4", increasing the seating capacity from 57 (41 second class, 16 first class) to 79 (60 second class, 19 first class). On the right can be seen the two-road car shed which was originally the terminus station until December 1896, when the line was extended to Dingle. The main depot was located at Seaforth Sands.

Photomatic

Modernisation of the existing stock was decided upon rather than the purchase of new stock, and the initial car to be so treated was No.29 in February 1945. The first complete modernised train entered service in 1947 and included motor car No.14, which is seen at the rear of a train at James Street on 2 June 1952.

J.C.Gillham

The second complete modernised train entered service later in 1947 but differed from the previous one in that larger windows and different door and window spacings were included in the modification work. The train is seen at Herculaneum Dock on 31 December 1955.

L.Collings

5

MERSEY & WIRRAL LINES
650V d.c.
3rd/4th rail (Mersey)
650V d.c. 3rd rail (Wirral)

Other railways electrified after the turn of the century all became part of British Railways from Nationalisation in 1948, but some systems have closed whilst others have been altered. The first steam-operated railway to be electrified was the Mersey Railway in 1903. The original stock was built by G.F.Milnes & Co. of Hadley in Shropshire. Motor car M28429 is seen at Birkenhead Park on 2 June 1952 in B.R. malachite green livery.

J.C.Gillham

Also photographed on 2 June 1952 seen leaving Birkenhead North is a train of Mersey Railway stock with first class motor car M28411 leading. The first class accommodation is recognised by the cream upright panels between the car windows. Note that the 3rd rail system only is used here, the section beyond Birkenhead Park not being electrified until 1938 by the L.M.S. During 1936, the Mersey Railway stock was modified so that it could operate on either the 3rd/4th rail or 3rd rail system.

J.H.Meredith

When signalling improvements were made to the Mersey Railway during 1921-2, additional stock was built by Cravens in 1923-4 so that improved services could be operated. The cars had elliptical roofs rather than clerestory common to the older stock, and one such train is seen at New Brighton in April 1955 mixed with older stock.
Real Photographs Co.

The building of the new stock for the electrification by the L.M.S. of the Wirral lines to West Kirby and New Brighton was shared between Metropolitan Cammell and Birmingham in 1938. Driving trailer M29276 is seen at Birkenhead Park in June 1952 in lined L.M.S. maroon livery.
J.C.Gillham

Replacement of the old Mersey Railway stock began in 1956 with stock similar to the 1938 design, operating on the 3rd rail system. Thus, prior to the new stock arriving, the existing 3rd/4th rail sections between Liverpool and Birkenhead Park/Rock Ferry were converted to 3rd rail only in December 1955, and the centre conductor rail subsequently removed. A train of old stock is seen here entering Birkenhead (Central) on the Rock Ferry branch, in the interim period.

Photomatic

The Mersey Railway replacement stock built in 1956 was again built by Metro-Cammell and Birmingham, and closely resembled the 1938 cars in appearance. However, there were several small detail differences, one being that the 'press to open' passenger door buttons were located on the car body, whereas on the 1938 cars they were on one of the door pairs. A 1956-built train arrives at Wallasey Village in June 1962 before the introduction of yellow warning panels.

G.F.Walker

A three-car train of 1956 stock at Birkenhead (Central). Note at the top lefthand corner the 'next train 3-cars' indicator, illuminating when such trains operate. Note the name 'Merseyrail' applied to the car sides since the Merseyside P.T.E. took over the service at the end of 1971, although B.R. continues to provide the rolling stock. The whole system (Liverpool to Rock Ferry/New Brighton/ West Kirby) was called the 'Wirral Line' on route maps. For a short time, the fronts of trains carried a 'smiling face', but this idea was soon dropped. During 1968-9, B.R. multiple units became classified by a numerical system, and although differences exist between the 1938- and 1956-built cars both are classified as Class 503.

Prior to the construction of the Liverpool loop single track tube tunnel, end communicating doors had to be built into the car ends as an M.O.T. requirement. This was started in 1972 and completed in 1976, the loop being opened in May 1978. A three-car 1938 stock train is seen at Birkenhead Park in June 1977. A further detail difference between the two types of stock is that the older cars have a beading line at waist level.

A three-car 1938 stock train approaching Rock Ferry, newly repainted in blue-grey livery, on 13 July 1979.

Birkenhead Park has been remodelled in recent years to provide only two through platforms and a double-faced bay platform, instead of four through platforms. A 1956 stock train approaches bound for the Liverpool loop, and shows the sharp curve west of the station and the redundant track area in the foreground.

Also in blue-grey livery, a three-car 1956-built train arriving at Birkenhead North.

Converted from spare Class 501 stock from the London area d.c. electric lines, a two-car battery unit stands next to the depot at Birkenhead North.

LIVERPOOL (L. & Y.) LINES
650V d.c. 3rd rail

All of the original stock of
the L. & Y. Liverpool-
Southport/Ormskirk lines had
been scrapped by 1946, save
for two cars retained for
parcels service, which
lasted until 1952. These
were replaced in 1953 by two
compartment stock cars
specially converted, the
compartment stock having
been built in 1926-7
primarily for the Ormskirk
branch. Parcels van M28497M
is seen here at Waterloo, as
converted from a driving
trailer; the other from a
motor car. Both survived
until 1964, although
passenger working with the
compartment stock ceased
much earlier.

J.H.Price

To replace the original
L. & Y. electric stock, new
units similar in appearance
to the 1938 Wirral cars were
built at Derby between 1939
and 1941. The most obvious
difference from the Wirral
stock was that the Derby-
built cars had air intakes
built onto the roof of the
motor cars for motor cooling.
This three-car train is seen
just after passing over
Duke Street level crossing
between Birkdale and
Southport on 27 March 1979.
This stock also carries the
'Merseyrail' lettering and
the whole system (Garston/
Liverpool-Southport/Ormskirk/
Kirkby) is now known as the
'Northern Line'. This stock
is now classed 502.

Up to 1977, a number of Class 502 units were formed of two cars (M-T or M-DT) and were coupled to a three-car unit at peak periods to form a five-car train. A two-car unit (M-T; the more rare of the two possible two-car formations) is seen in Southport sidings in June 1976, showing also the front of the three-car unit it is coupled to. Since the opening of the Liverpool link in May 1977 and the closure of Exchange station, the majority of two-car sets have been lengthened to three-cars.

A three-car Class 502 train arriving at Ainsdale in July 1976. An experimental hourly semi-fast service was introduced from May 1975 between Liverpool and Southport and for this the special 'Express' blind was used. The use of these was, and is, confined to the few peak-hour semi-fast workings.

Destination blind variations at Sandhills: An original blind which was a three-line destination for a two-line display is illustrated here. The other possibility was 'Via Marsh Lane to Ormskirk', but use of this ceased in 1957 when the route was closed to passenger traffic. Note this is one of the Class 502 units that is painted in blue-grey livery.

New blinds were required when the Kirkby and (later) Garston lines were electrified from May 1977 and January 1978 respectively, although not all Class 502 units received them immediately. A three-car train arrives sporting the 'new' Ormskirk; note that the leading car has had four window openings replaced by the non-opening type.

A new incorporation to the blinds was 'Formby', where there was only one peak hour short working. This was made redundant with the new timetable from May 1979 when the working was extended to Southport. The original 'down local' platform can be seen on the left; Sandhills once had four platforms, but was reduced to an island platform by October 1974.

One of the evening peak hour semi-fast workings to Southport, with a new blind. Compare this style with the other 'Express' blind shown on page 13.

New stock for the Northern Line commenced delivery at the end of September 1978 and comprised thirty three-car units of Class 507 built by B.R.E.L. at York. This is the first time that d.c. electric stock has been given the class and unit number together on the car ends; Unit 507.003 enters Southport in November 1978 when new into service.

New Northern Line Class 507 stock was delivered to Birkenhead North Wirral Line depot for commissioning and test running took place to and from West Kirby. Unit 507.006 in almost 'mint' condition is seen at Birkenhead North on 23 November 1978.

On 27 March 1979, the newest
Class 507 unit seen in
service was 507.020, seen
here at Birkdale.

When the Garston-Kirkby
service started in January
1978, some Wirral Line units
were loaned to the Northern
Line to supplement the
existing stock. A three-car
(1956 stock) train is seen
at Sandhills departing for
Kirkby. Note the
'Merseyrail' P.T.E. symbol,
and also that destination
boards are provided
- chalked in this case.

NORTH & SOUTH TYNESIDE
600V d.c. 3rd rail

The Tyneside electrification
started in 1904. In 1937,
new stock replaced the
original, built by
Metropolitan-Cammell and
comprising mainly
articulated two-car units.
Here a train of two twins
arrives at Percy Main on
10 June 1964. Electric
services were withdrawn in
1967, being replaced by
d.m.u.s. The Newcastle
Metro, at present under
construction, will take over
most of the North Tyneside
former electric lines.
G.F.Walker

In 1938, two parcels vans
were built, replacing those
originally built in 1904-8;
E29467E was photographed
outside South Gosforth car
depot.
J.H.Price

Additional stock for the
North Tyneside lines was
built in 1920-21, but was
made redundant by the new
stock illustrated on the
previous page. However,
it was modified and put to
work in 1938 on the newly
electrified South Tyneside
line to South Shields.
One such modified 1920-21
stock train enters Tyne
Dock in August 1954.
                J.Wyndham

Replacement of the 1920-21
stock on South Tyneside
services came in 1955, when
new stock built to B.R.
standard design at Eastleigh
was delivered. Seen here at
Newcastle is a two-car unit
with motor car E65318
leading.
                J.H.Price

The new B.R. stock for the South Shields line had a relatively short life in the Newcastle area, the service being replaced by d.m.u.s in 1962. The stock was transferred to Southern Region and modified, where they are still at work today, being classified 2-EPB. The modifications included replacing the destination and headlight panels with a two-track number blind. This two-car train is arriving at Newcastle in August 1960.

J.Tilley

In their new role as 2-EPB units on Southern Region, Unit 5790 departs London Bridge in green livery. These can be distinguished from the main standard design 2-EPB units in that the luggage compartment has a small window, and the numerical headcode panels are slightly smaller.

G.F.Walker

LANCASTER-MORECAMBE/HEYSHAM
6,600V a.c. overhead

The original stock was
withdrawn in February 1951
and from 17 August 1953
services were provided with
converted stock. This was
of 1914 origin and comprised
three-car units that were
built by Metropolitan
Carriage Wagon & Finance Co.
with Siemens equipment for
the Willesden-Earls Court/
Kew Bridge services. They
were withdrawn from service
in 1940 and stored until
converted at Wolverton for
their new role. The
conversion work included
major alterations to the
driving ends of the motor
cars to operate at 6,600V a.c.
overhead. One of the
converted sets enters
Lancaster on 12 June 1964.
                G.F.Walker

The 1914 Siemens sets
comprised four three-car
sets, but only three were
initially provided for the
service from 1953. However,
in 1957, the fourth set
entered service and differed
from the other three in that
the louvres on the motor car
(M28222M) were left in the
original position.
          Real Photographs Co.

The driving trailers were virtually unaltered in appearance; this is M29023M leaving Lancaster on 12 June 1964.

G.F.Walker

Some of these sets obtained yellow warning patches before being withdrawn in 1967, as illustrated by driving trailer M29021M.

J.H.Price

MANCHESTER-BURY
1,200V d.c. 3rd rail

After experiments with
overhead electrification
from 1913 at 3,500V d.c.
between Bury and Holcombe
Brook, the Manchester-Bury
line was electrified at
1,200V d.c. 3rd rail in
1916. The Holcombe Brook
branch was similarly
converted in 1918. For the
3rd rail system, new rolling
stock was built at Newton
Heath between 1915 and 1921.
In this scene, a train of
original stock approaches
Crumpsall.

F.W.Ivey

Trains at Bury with 1920-built
car leading (left) and 1916-
built car (right), although
identical in appearance.
Note that the latter does not
have an 'M' suffix to the car
number.

J.H.Price

The newest motor car of the original stock to be built (1921), photographed at Bury, numbered M28537M. It was numbered 3539 when new, 14609 by the L.M.S. and later 28537 by the same company. The 'M' prefix and suffix was applied after Nationalisation.

Real Photographs Co.

Transition stage on the Manchester-Bury line with both old and new stocks in service.

F.W.Ivey

New stock was built at Wolverton for the Bury line in 1959-60. Two two-car units with motor car M65446 leading are seen when newly in service.

F.W.Ivey

Two-car trains are operated during off-peak periods as illustrated here by driving trailer M77182 leading into a deserted Woodlands Road station.

Livery variations on Manchester-Bury stock; after the lined green, the yellow patch was added. This semi-derelict driving trailer is at the back of Bury depot in July 1976, on accommodation bogies.

A number of cars made redundant from service reductions have not seen service for a number of years and still sport the original lined green livery, but with a yellow warning patch. During 1978-9 these cars (with others) were transferred to Croxley Green depot in the London area for under-cover storage.

In B.R. blue livery, but with
a yellow warning patch,
another stored driving
trailer stands in Bury depot
yard on a typical wet
Manchester day in July 1976.

When all-yellow front ends
were introduced, this
extended round the cab back
to the cab door as shown on
M65443, also out of service
at Bury in July 1976.

L.M.R. LONDON AREA
630V d.c. 3rd/4th rail

Following the four trains of 1914 Siemens stock that were later put to work on the Lancaster-Morecambe/Heysham line, many cars similar in appearance were built by Metropolitan Carriage Wagon & Finance Co. (motors) and Wolverton (driving trailers and trailers) with Oerlikon equipment between 1915 and 1923. Motor M28253M leads this six-car train at Bollo Lane, Acton, on 24 August 1957 having diverged from the (now) freight-only line to Kew Bridge.

J.C.Gillham

Oerlikon motor M28264M departing from the 'old' Euston before the rebuild for a.c. electrification. This car was one of twelve motor cars which had been fitted with a glass panel next to the sliding passenger doors.

G.F.Walker

An Oerlikon driving trailer leading a three-car train in one of the platforms at Willesden Junction. This bay platform has since been removed and only the one on the right now remains, capable of accommodating a three-car train only.

F.W.Ivey

Richmond station in the early 1950s with motor car 28232 still in lined L.M.S. maroon, and one of the cars with a glass panel next to the passenger door. The trailer M29730 is one which had recently been repainted in the then-new B.R. malachite green livery. The suffix 'M' was not applied to cars until the end of 1951, which was used to denote the region of origin; the prefix denoted the region the stock was allocated to.

F.W.Ivey

Branch line services from Watford: M28231M leads a three-car Oerlikon train at Rickmansworth (Church Street) on the last day of service on that branch, 2 March 1952.

J.H.Meredith

Croxley Green branch train of Oerlikon stock at Watford Junction on 13 April 1957. This branch, although surviving, now only operates at peak periods on Mondays to Fridays. At one time through peak services used to operate direct from Croxley Green to Broad Street. Oerlikon stock was withdrawn from service gradually from about 1955, but lasted on the Croxley Green branch until 28 April 1960.

Capital Transport

Additional stock was delivered between 1927 and 1932 of the compartment type with G.E.C. equipment and doors to each seating bay. At the same time identical cars were built for the Liverpool to Ormskirk/ Southport lines. The 1927 motor cars were built by Metropolitan Carriage Wagon & Finance Co., trailers by Clayton Wagons of Lincoln and driving trailers by the Midland Railway Carriage Works at Birmingham. The 1929-32 batches were all built at Wolverton. Motor M28022M leads a Broad Street train at Dalston Junction, despite the destination blind which has already been changed for the return journey, on 9 November 1957.
J.C.Gillham

A compartment G.E.C. stock three-car train at Kilburn High Road on 20 March 1957 with a driving trailer at the rear.
Capital Transport

The G.E.C. stock outlasted
the Oerlikon by three years,
the last being withdrawn in
1963, but not before some
had been painted with a
yellow warning panel as
seen on this six-car train
at Acton on 12 April 1963.
G.F.Walker

Oerlikon and G.E.C. stock
could be operated in service
together, but rarely did so.
One of these rare occasions
is depicted at Gunnersbury
on 8 June 1957 as a mixed
stock train for Broad Street
arrives.
L.A.Mack

New stock for the L.M.R. London area d.c. lines was built by B.R. at Eastleigh from 1957 and initially replaced the Oerlikon stock, followed later by the G.E.C. stock. A train of Oerlikon stock is seen next to a 1957-built train in the sidings at Harrow and Wealdstone.

F.W.Ivey

A 1957 stock train in original livery, after the application of a yellow warning panel but before it was decided to apply the whole front end. This three-car train is entering Acton (Central) on 2 June 1962, having worked 'wrong line' due to engineering work.

G.F.Walker

One of the first livery alterations was repainting in lined green, as illustrated by this three-car train entering Acton (Central).

G.F.Walker

Since the Colne Junction-Croxley Green Junction direct connection was closed in 1966 all stock now faces the same way - motor cars south and driving trailers north; a three-car train approaches Stonebridge Park in August 1978, clattering over the recently installed connection to the new L.T. depot.

A six-car train of two
three-car units at Hatch End,
with driving trailer M75135
leading. Like all other B.R.
suburban stock in recent
years, the standard livery
has been blue with all-yellow
front cab ends.

A stranger to the L.M.R.: on
2 August 1970, the stock
operating the L.M.R. London
area d.c. lines was converted
from 3rd/4th rail to 3rd rail,
but the 4th rail was retained
where L.T. trains continued
to operate. However,
a special train of S.R. 4-COR
stock is seen on a L.C.G.B.
rail tour, made possible
because of the 3rd rail
conversion, at Croxley Green
on 8 November 1970. The
redundant 4th rail had not
been removed by that time.
                    K.Gunner

MANCHESTER-ALTRINCHAM
1,500V d.c. overhead

The first use of e.m.u.s for
1,500V d.c. overhead
electrification was in 1931
on the M.S.J. & A. Railway
between Manchester (London
Road), now known as
Piccadilly, and Altrincham.
The rolling stock was built
by Metropolitan Cammell.
Additional trailers were
obtained in 1939 comprising
two newly-built cars from
Wolverton, five transferred
from the Liverpool area plus
one from the London area.
This six-car train is seen
at Altrincham in B.R.
green livery.

F.W.Ivey

Most of the original stock
lasted until withdrawal in
1971, when it was decided
to convert the line to
operate at 25kv a.c. overhead
and integrate the service with
that between Manchester and
Crewe. From 1968 motor cars
had their diamond-shaped d.c.
pantographs replaced by the
smaller a.c. type, as depicted
on this train leaving
Altrincham on 10 April 1971.
Note also that as late as 1971
only a yellow patch is still
on the motor car end.

J.C.Gillham

Close-up of motor car M28581, some of which were painted in blue livery with all-yellow front ends prior to withdrawal. Note also that the 'M' suffix has been omitted from the car number and that plain panels have replaced the side ventilator louvres.
J.C.Gillham

Another train in blue livery on 10 April 1971, the leading motor being correctly numbered with the 'M' prefix and suffix. The a.c. service started on Monday 3 May 1971, the last use of this stock being on 30 April 1971.
J.C.Gillham

MANCHESTER-GLOSSOP/HADFIELD
1,500V d.c. overhead

A six-car train at Manchester (Piccadilly) in July 1975 shows clearly that headlight codes are still used on this line.

Recently, three-car trains only are operated, although the weekday service intervals have been improved to operate every thirty minutes or every fifteen minutes at peak periods. A three-car train arrives at Godley which, until recently, was named Godley Junction.

A three-car set arriving at
Dinting in July 1979 shows
the severe curves at this
station and the redundant
platform on the Hadfield to
Glossop side of the triangle.
                    J.P.Herting

Driving trailer M59603M leads
a three-car train into
Broadbottom. This station
was named the mouthful
'Broadbottom for Mottram and
Charlesworth' until recently.

EASTERN REGION:
Lines from Liverpool Street

Electric services on the Liverpool Street-Shenfield section of Eastern Region started in September 1949 and operated at 1,500V d.c. overhead. At first the stock carried no unit numbers, as illustrated by this up train entering Stratford on 25 March 1950.
J.H.Meredith

Trains were later given unit numbers on discs attached to the front of trains as shown by Unit 89 at Ilford on 25 March 1956, after repaint and overhaul. Note the headlight codes fitted and the similarity in design with the Manchester-Glossop/Hadfield stock, on which the latter was based.
L.Collings

Most of the d.c. sections were converted to a.c. in 1960 and the stock subsequently rebuilt for a.c. operation. The conversion work entailed moving the pantograph from the leading end of the motor car onto the trailer. In April 1979, a train is seen approaching Rochford on an empty working to Southend, and shows clearly where the lower roof line used to be. This stock is now known as Class 306.

When the stock was converted from d.c. to a.c. operation, the headlight codes were replaced by two-track numerical blinds and units were renumbered with an 'O' prefix. Driving trailer E65611 leads a nine-car train into Gidea Park. Note that this stock has sliding doors and passenger door control buttons. Being built in 1949, this stock is due for replacement in 1980-1 by new Class 315 units.

Also electrified at 1,500V d.c. overhead was the line from Shenfield to Southend (Victoria) and for this service new stock was built by B.R. at Eastleigh. These were given unit numbers from the start, but were suffixed 'S' to distinguish them from the local stock. Unit 12S is seen at Eastleigh on 26 May 1956 when new.

L.A.Mack

The Southend line too was converted to a.c. operation and the stock subsequently converted. The first unit to be dealt with was 03S in 1959, and this was renumbered 103 to avoid confusion with the 1949 Shenfield local stock.

Conversion to a.c. (6.25kv inner area and 25kv outer area) included moving the pantograph and the guard's compartment to the driving trailer, one of which (E75012) is seen at Southend (Victoria) in May 1979. All units were renumbered 1xx when converted.

Other conversion work on this stock, known as Class 307, included replacing the headlight code panel with headcode number blinds. Here, Unit 131 approaches Romford.

The new front end design was used on the Class 308/1 units, nine of which were delivered in 1961 for the Colchester-Clacton and Walton section, and a further 24 in 1962 for through services. These were built for a.c. operation when new, as was all other Eastern Region electric stock. Unit 137 is seen at Thorpe-le-Soken in April 1979; note the unofficial chalked 'daffodil' above the B.R. emblem.

Prior to the Class 308/1 units there were nineteen units in Class 305/2, built at Doncaster in 1959-60 for the Liverpool Street-Bishops Stortford/Hertford East line. However, as Unit 506 at Shenfield illustrates, the units tend to be used on any section according to stock availability and requirements.

Built at York in 1960 for suburban services from Liverpool Street to Chingford/Enfield Town were 52 units of Class 305/1. These differ from 305/2 in that the interior layout is of an open arrangement instead of compartments; part of Unit 437 awaits attention in Ilford depot.

Following on in the numbering from the Class 305/1 units were three Class 308/3, built at York in 1961 for the same work as the Class 305/1. Having different equipment, the design is the same as the Class 305/1 stock. Unit 455 stands in Stratford yard in May 1962 and shows that some E.R. units had their car numbers prefixed 'GE' at that date.

G.F.Walker

For the express services between Liverpool Street and Clacton/Walton, new stock gangwayed throughout was built at York in 1962-3. This stock is now known as Class 309 and was divided into three groups. Seen at Liverpool Street is Class 309/3 Unit 624 in its original livery of lined maroon, with 'wrap-around' cab windows.

B.R.Hardy Collection

Seen arriving at Thorpe-le-Soken in April 1979 is Unit 622 of Class 309/3, with modified cab windows. Clacton/Walton trains usually divide at this station, the front unit going on to Walton and the rear to Clacton.

In the opposite direction at Thorpe-le-Soken, trains from Clacton normally arrive first as depicted by Class 309/2 Unit 611 in April 1979, still with the original design of cab windows. Class 309/2 units differ from the 309/3 in that a buffet car is included in the formation.

At the other end of Thorpe-le-Soken, the Walton portion arrives cautiously to couple to the portion from Clacton, already in the platform. Note the spasmodic use of headcodes on this stock.

To enable the Clacton portions to have more accommodation than the Walton trains, Class 309/1 two-car units were also built, the pantograph on this class being located on the driving motor car. Driving motor car E61943 of Unit 604 stands in isolation outside Clacton depot while its driving trailer receives attention in the depot on 12 April 1979.

In 1974, four units of Class 309/1 were lengthened to four-car units by the addition of two intermediate trailers and were reclassified 309/4, making four sub-classes in all; Unit 605 stands at Liverpool Street in 1975, in the days of guaranteed correct headcodes.

Additional semi-fast four-car units were built at York in 1975 for improved services from Liverpool Street. They are similar in appearance to the Class 310 units built in 1965 for the L.M.R. a.c. lines (see pages 59-60) except for flat cab windows. These units are classified 312/1 and were the first units to be delivered with the class and unit number side by side on the front. Confusion soon arose, as the units were referred to with the last three numbers, clashing with the Class 307 units. Paper stickers were thus used to eliminate confusion on an unofficial basis, as displayed by 312.104 (784) at Colchester on 12 April 1979.

The Class 312/1 units tend to travel over most of the lines into Essex from Liverpool Street. Here, 312.109 (789) stands at Southend (Victoria). All 312/1 units have now been renumbered incorporating the once unofficial numbers, so that this example is now 312.789.

EASTERN REGION
Lines from Fenchurch Street
6.25/25kv a.c. overhead

The other Eastern Region a.c.
overhead electrification
scheme of the early 1960s was
that from Fenchurch Street to
Shoeburyness including the
Tilbury line. For this,
112 four-car units were built
at Doncaster and York between
1958 and 1960 and are now
classified Class 302;
Unit 289 leads a twelve-car
train at Barking in July 1976.

The use of headcode blinds and
destinations has now ceased on
the L.T. & S. Class 302 stock,
as illustrated by Unit 258
arriving at Pitsea.

Apart from some peak hour workings which avoid Tilbury (Riverside), all trains have to reverse in the station to whichever destination they are bound for. Unit 300 arrives from London (left) and will depart to the right for stations to Southend.

In addition to the Class 302 units on the L.T. & S. line, a further nine four-car units were built at York in 1961 and each included a luggage van in the place of a passenger car. These are now classified 308/2. Four units with luggage vans were rebuilt at Wolverton in 1971 into passenger vehicles, and the relevant units reclassified 308/4. Class 308/4 Unit 316 heads a Shoeburyness train through Woodgrange Park on 19 May 1979 when, due to engineering work, Fenchurch Street services were diverted via Stratford. This station is served only by d.m.u. trains on the Kentish Town-Barking line.

The 'odd' unit on the
L.T. & S. line is 244, seen
at Fenchurch Street. The
leading car is a replacement
for a collision-damaged
driving trailer which was
scrapped. The replacement
was obtained from the
Manchester-Bury line.

K.Gunner

The same car before
conversion to a.c. working,
in Wolverton yard on
18 October 1970.

K.Gunner

Seen leaving Stratford on 24 June 1961 are Class 302 units on loan to the G.E. lines from Liverpool Street while their stock was being converted from d.c. to a.c.
G.F.Walker

Some of the Class 302 units are still allocated to Ilford and are, in the main, the earlier units. 208 departs from Gidea Park on 12 April 1979.

EASTERN REGION
Great Northern Suburban
25kv a.c. overhead

The first stage of E.R.'s Great Northern suburban electrification scheme opened on 16 August 1976, when services were provided between Drayton Park and (initially) Old Street and (later) Moorgate over the former L.T. Underground line. This section operates at 750V d.c. 3rd rail and 25kv a.c. beyond. The changeover of voltages is effected in Drayton Park platform. On the first morning of operation, Class 313 Unit 313.013 arrives at Drayton Park. The disused L.T. depot on the right has since been demolished.

Services to Hertford North and Welwyn followed in November 1976, and on this section Unit 313.006 leads a six-car train into Cuffley in April 1978. The first ten Class 313 units have raised roof sections for interior ventilator fans. The Class 313 units were built at York and were the first suburban e.m.u.s to be painted blue-grey when new. Note that by this date headcodes had been discontinued.

Unit 313.063 at Hatfield; these blinds have been superseded by the black lettering on white background type.

For the Great Northern electrification, additional battery units were converted from surplus L.M.R. London area Class 501 motor cars, as seen in Hornsey depot. A further two pairs are at present being converted at Doncaster for the Midland Suburban electrification scheme.

For the outer suburban G.N. electrification, 26 Class 312/0 units were built at York, almost identical to the Class 312/1 ones operating from Liverpool Street. The main noticeable differences are the blue-grey livery and black painted cab window surrounds. 312.019 stands at Royston, the outer terminus for electric trains.

Although built with headcode panels, these were obsolete from the start, as 312.022 shows at Finsbury Park in April 1978.

LONDON MIDLAND REGION
Western a.c. lines
25kv a.c. overhead

The Manchester-Crewe section
of the L.M.R. main line was
electrified from
12 September 1960 and for
this line, fifteen units
(Class 304/1) were built
at Wolverton. Unit 012
arrives at Hartford on the
Liverpool-Crewe section,
which was electrified from
1 January 1962 and shows
that the stock soon became
mixed between the two
sections.

G.F.Walker

For the Liverpool-Crewe
section, further similar
units were built at Wolverton
and became Class 304/2, the
main difference being altered
seating accommodation. In
August 1977, Class 304/2 Unit
034 stands at Walsall.

The third group within the 304 Class was the 304/3, built in 1961; Unit 043 leaves Allerton for Liverpool on 13 July 1979.

The Class 304 units are rarely to be seen south of Rugby, but one such occasion was on 14 May 1966 when Class 304/3 unit was photographed on a northbound eight-car train near Wembley (Central).

G.F.Walker

For outer suburban services from Euston, fifty Class 310 units were built at Derby between 1965-7; Unit 075 in original condition departs Euston for Rugby on 12 August 1966.

G.F.Walker

Apart for the cab ends being painted all-yellow, Class 310 units are virtually unaltered in appearance since new; Unit 085 leads a train into Tring in August 1978.

Unit 078 arriving at Lichfield (Trent Valley) on 4 November 1978 on the infrequent local service between Stafford and Nuneaton.

Class 310 a.c. stock rarely works local services to Watford Junction, this being provided by the d.c. stock which stops at all stations. However, one such working is seen in the a.c. bay platform on 16 August 1977. These can only stop at Bushey, Harrow, Wembley and Queens Park, however.

Class 310 units work regularly to Manchester (Piccadilly), but only very occasionally on the Altrincham branch; Unit 074 is seen here at Piccadilly on 17 September 1976 on an Altrincham-Alderley Edge service.

For local services between Birmingham (New Street) and Birmingham International (the National Exhibition Centre), four Class 312/2 units were built at York in 1976. Although similar to the 310 units of the mid-1960s, they have been fitted with flat cab windows; 312.202 is seen at Birmingham International.

SCOTTISH REGION
6.25/25kv a.c. overhead

All original 91 three-car units for the North and South Clydeside electrification were built between 1959 and 1961 by Pressed Steel. Although initially the units were kept on their respective lines (001-056 North and 057-091 South), interchange of units became common. At first, no unit numbers were carried, as illustrated by a train approaching Balloch on 13 May 1962.

G.F.Walker

On the North side, a small part of the system runs underground in double-track tunnels; Unit 052 leads a Bridgeton train at High Street. The destination blind has already been altered for the return journey.

On the South side, Unit 040 photographed at Kirkhill; this stock is known as Class 303.

When the 'South Bank' lines to Wemyss Bay and Gourock were electrified in 1967, new stock similar to the Class 303 units were built by Cravens. These too became intermixed on all the systems and Class 311 Unit 107 is seen leaving Dalmuir on the North side system. Note the Greater Glasgow P.T.E. symbol 'GG'.

Both Class 303 and 311 units are being repainted in blue-grey. Class 311 Unit 110 is seen at Springburn on 28 June 1978 in the new livery, also having the class number on the unit front.

New stock for Glasgow's Argyle line is being built by B.R.E.L. at York, classed 314. At the 'new' Bridgeton station on 4 June 1979, Units 314.201 and 314.202 stand after a press run from Glasgow (Central) Low Level after the 'switching-on' ceremony. At the time of writing, public services are expected to commence on 15 October 1979.

A.Stirling